Text: Kim Loughran
Editor: Christopher Westhorp
Picture editor: Erik G Svensson
Design: Patric Leo
Layout: Petra Ahston Inkapööl
Repro: Linjepunkt, Falun
Printing: Livonia, Lettland 2014
www.maxstrom.se
ISBN: 978-91-7126-266-0

The Year in Stockholm

KIM LOUGHRAN

BOKFÖRLAGET MAX STRÖM

(Page 1) Stadshuset (City Hall) and Lake Mälaren. The gilt top of the tapered tower is topped by Sweden's national emblem, the Three Crowns.

The archipelago city. The Royal Palace in the centre seems to watch over all sides of its surrounding city. To the right is Gamla Stan (Old Town) with the Riksdagshuset (Parliament) in the left foreground. The extended archipelago, reaching to the Baltic Sea on the horizon, is made up of 30,000 islands. The city has a metro-area population of two millions. Capital of Scandinavia? Stockholm is central to the region and is its biggest, fastest growing city.

STOCKHOLM IS ONE third water, one third green space, one third built-up. All scenically spread over 14 islands. An archipelago is just a short steamboat ride from the Royal Palace.

It is a prosperous town. You get efficiency and well-oiled infrastructure. This is one of the fastest-growing cities in Europe, with a healthy birth rate. This is a well-ordered, consensus-run modern city with people moving in the same gear.

For all that growth, the city consistently scores high on rankings such as the UN Habitat Prosperity Index, based on productivity, good governance, education, entrepreneurship, city planning, crime, air quality, IT, infrastructure, equity and social inclusion, health, environmental sustainability, quality of life – I could go on. Stockholm was the first city to be nominated EU Green Capital of the Year, for its consistent record of high environmental standards and a commitment to sustainable development.

No big-city attitude? Perhaps in the distances between people, a legacy of dire winters. Being polite means minding your own business. Stockholmers love their independence and the city has a high proportion of single-person households (almost two thirds).

Stockholmers are proudly cosmopolitan, eager to accept ideas, safe in their strength as multi-lingual trendsetters. With the best coffee on the planet. (Shared prize.)

Stockholmers inhabit a delta and an archipelago, with the pulsing heart of a medieval town, but loyal to their own districts, relaxed about being aloof but glad to share friendship with visitors.

You'll notice an architectural style – best seen in our Stadshuset (City Hall) – that draws on Roman and Greek classicism with Italianate and Nordic Romanticism: stripped and bright, with a yellow stucco base catching any available light.

There is a 19th-century Europe feel in parts of the city, notably posh Östermalm and vibrant Södermalm. In the mix is a food culture that relishes trends, continually refining the Swedish blend.

A sub-culture of sharp geeks has given the city the sharpest conglomerate of IT business successes in northern Europe. Stockholm is consistently among the world's top three most wired-up cities. Skype and Spotify were developed by Stockholmers. Game-creators are working overtime as I write. Music throbs in the clubs and pubs. Stockholmers have a buzz. Don't worry, you'll find it.

You will need a few survival tricks: Learn the difference between regular and cowboy taxis with the spinning meters. In a bar, check the pricelist before you order drinks. Service is always included.

Don't dawdle at check-out counters. Walk to the right. Be on time! Stockholmers are no-fuss people.

Welcome.

PS
Now you're here, why not stay? According to Global Age Watch, Sweden is the best country in the world to grow old in. And Stockholm is its heart.

The narrow lanes of Gamla Stan, the medieval Old Town, echo with history, from the days of a vicious bloodbath at the hands of the Danes, to the commercial Hanseatic period when German was the required language for city councillors and, in the last century, the more tranquil era of artists and poets.

3.1

JAN

The freshwater bay of Riddarfjärden freezes every few years, broadening the popular waterside promenade. The street to the left, Norr Mälarstrand, was where 1920s film stars chose their expensive apartments.

(Overleaf) The church spires of the city, hemmed in by communications spires on the outskirts. Stockholm's oldest building is the church at Riddarholmen (centre). To the right of there is a city hub: the Slussen locks. The winter sun is setting, through a plume of water vapour, on the European capital city with the cleanest air.

THE WORD 'REALISATION' appears in shop windows. In Swedish, it means Sale! Stockholmers are looking at their best now, decked out in fashions from the January sales. Flair and quality mark the Stockholm style, prim and proper but quality-minded. Fashion is a dynamic local industry, in design and retail. Successful producers include a couple of ersatz American ones: Gant and Peak Performance – now respectful curators of distinct styles. In retail, the Persson family's H&M is a global player. Stockholm has an affinity with New York, and the hip Hornstull area is known as our Brooklyn.

The weather? January weather is stable. Either mild at zero or crisp at minus five. Pavement snow gets tramped into ice; occasional thaws slicken the surface, so pedestrians had better watch out. We walk like penguins. Looking up. Deadly icicles can be pointing at you. The roofs of the city's rich trove of old apartment buildings are

The Ericsson Globe arena squats, like a friendly mothership, on the southern skyline.

Free skating lessons at the Kungsträdgården rink. The park is bordered on one side by the country's mightiest banks.

(Overleaf) No such thing as a free lunch? Archipelago avians munch on maize, wheat and barley snacks, meanwhile providing one of the city's attractions. Over 100 species have been recorded at the "lunch counter", making the spot a favourite for birders. The prime minister's office glows bright sandy-brown in the distance. Parliament is on the left, the Royal Opera House on the right.

OPERAKALLAREN

not insulated as before. The boom in re-vamping attics into fancy digs has robbed buildings of a layer of insulation, causing snow-dumps and icicles. Watch out.

The scent of fir is in the air. Discarded Christmas trees are piling up on 20 designated city squares in what look like funeral pyres. But that's exactly what is happening: each incinerated Christmas tree can power four loads of washing. A hundred firs can power a normal refrigerator/freezer for a year.

Recycling is part of a city's hygiene and Stockholmers like things clean. Anti-littering laws are tough (except for cigarette butts).

The EU has praised Stockholm for its "innovative integrated waste system" and when the EU chose Stockholm as its first-ever Green city in 2010.

It is still dark, but clear January skies give us delightful dawns. Because darkness is still the main theme, January is an excellent time for ghostbusting. The Stockholm Bloodbath of 1520 and the traditionally riotous inn culture have left Gamla Stan, the Old Town, teeming with night spectres.

Across Strömbro Bridge, the Kungsträdgården Park skating rink is offering free lessons. Come to skate, and stay to watch families teaching little children.

Turn back towards the harbour and to your right you will see the Kungliga Operan (Royal Opera House) complex. Behind the glass veranda of the Operakällaren restaurant rises the Royal Opera House itself, the workplace for years of Sweden's greatest soprano, Birgit Nilsson. The Royal Opera was site of a regicide, the assassination of King Gustaf III – ironically, the man who commissioned it.

From a cheap seat on the third tier of the Royal Opera's beautiful baroque interior, for the price of a couple of beers, you will not see much action but will hear world-class music. If you are under 26, it's cheaper than a movie ticket.

Looking like an anonymous, fat troll with a hat on, the *semla* starts the Swedish bakery year with a crescendo. Resistance is

A Lenten Bun, or *semla*, is too delicious to schedule. Traditionally, you're meant to wait until the last Tuesday before the Christian fasting season of Lent before munching into one. Stockholm's cake shops blatantly ignore the convention and start selling from the first week of the year.

(Overleaf) The island of Skeppsholmen. For three centuries, this was naval headquarters – until the advent of the atom bomb prompted the decentralisation of military headquarters. In the distance to the left, on another island, is Nordiska Museet (Nordic Museum).

futile. These are Lenten Buns and a treat for Christians facing 40 days of Lenten fasting before Easter. Today, more Stockholmers might observe Ramadan than Lenten customs but everyone loves a *semla*. Officially, the *semla* period starts later, around Fettisdagen (Fat Tuesday, or Shrove Tuesday), in February or March, but there's nothing official about the enjoyment of whipped cream and marzipan atop a light, cardamom-flavoured bun.

Talking of over-indulging; the gyms are full of Christmas victims (no names, please) making New Year resolutions. January is meant to hurt a little after all that fun at New Year.

NORDENS LJUS
Rederi Stockholms Ström

Enjoying Ekoparken, the world's first urban national park. The island park has forest landscapes, restaurants, marinas, ferry piers, art museums, an outdoor history park with animals, and a dedicated ABBA museum, next to an amusement park. But in the winter, skiers have it almost to themselves. Walkers love its 20 kilometres of waterside paths.

FEB

THE YEAR'S FIRST school vacation. It's a week called *sportlov*, "sports break", a chance for the nation's youth to take exercise on snowy slopes, breathing the clean air of the backwoods.

Yeah, right. More and more, it's a chance to hang out with parents on the sands of Dubai, Phuket or Fort Lauderdale.

But for city kids uninterested in travel, the manmade mountain with real artificial snow at Hammarby *skidbacke* (93.5m-high) is a magnet. Four slopes have been built on rubble left over from the construction of the Ericsson Globe, the world's biggest hemispherical building.

The Globe is the most central major arena, although it is smaller and older than the 65,000-seat Friends Arena in Solna district and not big enough for football. Stockholm has a glut of arenas and scarcely enough sports or other entertainment to make them profitable. The Friends Arena is the home ground of the black-and-yellow AIK football side, known as Gnaget, roughly translated as "rats who never give in".

Their cross-town rivals are Djurgården IF, known as the Iron Stoves (*järnkaminerna*), either because they were tough or because the blue striped sweaters looked like the front of a stove.

The third city side is Hammarby IF, who don't have an exciting nickname but do have a more colourful fan base. All three clubs have multi-sport programmes, fielding top-echelon sides in a range of summer and winter sports. No other city in Sweden has this. All three clubs are (formally) membership-owned. Hammarby and Djurgården IF share the Tele2 Arena, neighbouring the Ericsson Globe.

The traditional favourite winter sport is *bandy*, a ball game played with sticks and on skates on frozen football pitches. The plastic ball is about the same size as an orange and coloured orange for visibility. It originated in 10th-century Russian monasteries and came to Sweden, via England, 300 years ago as a game

Most ferry services shut down for the winter but a few operate year-round. The ice is kept open by constant traffic. In the summer, about 50 ferries dart back and forth between island jetties.

(Overleaf) Gliding gracefully down the ski slope at Hammarby. The hill was shaped from rock and soil excavated from building sites when the central business district was largely rebuilt in the mid-20th century.

KONTOR

Not dressed for the season and with his lyre useless under snow, Apollo gazes longingly towards the heated interior of Drottningholm Palace, official residence of King Carl Gustaf and Queen Silvia.

(Overleaf) Steering straight at the Stadshuset (City Hall), this skater is making the most of Riddarfjärden Bay's pristine, freshwater ice. Stockholmers are quick to make the most of winter ice. Skates are sharpened at shoe repair shops.

for aristocrats. Stockholm's glamour team is Hammarby, more successful in *bandy* than football (a summer sport in Sweden). Never to be forgotten is the muffled sound of gloved hands of *bandy* fans applauding in below-zero temperatures. In Stockholm, Bandy Day is on or around 10 February and the big games are at the Zinkensdamm outdoor rink.

The northernmost indigenous people of Europe, the Sami, inhabit a region the size of Sweden but spread over parts of Sweden, Norway, Finland and the Kola Peninsula of Russia. Some say the biggest concentration of Sami is in Stockholm. Whatever the truth in that, their national day on 6 February is celebrated at Skansen, the museum park. The Sami are a hardy, outdoor people, so you'll be offered treats such as coffee with salty reindeer bouillon, and *palt,* dumplings made with pig's blood. You'll see a few reindeer at Skansen but not enough to confirm that Sweden's reindeer always run counter-clockwise when corralled. This is known as the Coriolis effect, due to the Earth's rotation, also affecting bath water.

Skansen is on Djurgården Island, and close to the Liljevalchs art gallery. The gallery's spring show, generally in February, is Sweden's highest-profile amateur art show. Anyone can enter a work and thousands do. Like reality TV, only without the public shaming.

February has a bad rep. It's peak month for whatever flu has been going around. Feel free to complain. February doesn't feel like raising your hopes, thank you very much. But see the snowdrops in the parks and hear the chirping in the trees. It's the *talgoxe* (great tit) or the *blåmes* (tit), back in town to compete with gulls for scraps, smaller but cleverer.

Suddenly, the days are much longer.

MAR

The sun is still low in March but lasts longer. The narrow waterway of Pålsundet links the old prison island of Långholmen to the southern island of Södermalm. Soon pleasure boats will be jostling for parking space.

BLUE HORIZONS APPEAR. Crocuses are popping up in parks and gardens. The sun sets later and hope exists. Hearts rise, but the temperature is volatile.

Song swans are nesting along Djurgården Park's bays and inlets. Larks – the woodlark (*Lullula arborea*) and the skylark (*Alauda arvensis*) – are back in the treetops of Humlegården Park, singing to the toddlers in the park's play space. But especially serenaded is the statue of Carl Linnaeus in the middle of the park – the 18th-century bird-lover who taught the world to know its flora and fauna.

Snowdrops (*Galanthus*). No braver flower exists. March brings blossoms, buds and flowers to countless corners and backyards. The city's seven nature reserves and the Ekoparken are preparing for spring.

The linden trees and Norway maples along the boulevards (Karlavägen, Ringvägen, Sveavägen, Narvavägen, Strandvägen, Valhallavägen) are getting a trim. Along the Klara Mälarstrand canal, the weeping willows (*Salix x pendulina*, "Elegantissima") are slowly sprouting their thin leaves.

The snow is mostly gone. Pavements are crunchy with gravel spread over the winter ice. Perky CityCat and Jespo machines spin along the pavements brushing away dust and sucking up gravel for cleaning and re-use.

March is the last chance for tour skating: to skim across the thin ice of the archipelago where the smoothest ice is closest to open water. To join a tour skating club, you have to leap fully clothed into freezing water and claw your way out.

Tour skating in the archipelago is spectacular but the waters of Lake Mälaren also provide prime ice. Local conditions are as good as, or better than, anywhere in the world. If it's still below zero, you'll see people practising at the track at Östermalm's Idrottplats sports arena. The ice skating and sailing club, SSSK, is Scandinavia's oldest. Interestingly, the temperature in the city can be above zero and yet in parts of the archipelago and the lake, ice will still hold.

But this is a month for turning over the soil in garden allotments. The city has 45 designated clusters of these *kolonilotter* with their huts and tiny gardens, scattered around the metro area, some in

impossibly small pockets. The lots were once far from crowded city streets – now they're surrounded.

Out-of-towners find Stockholmers boastful, foreigners can find us detached. Are Stockholmers happy? Happiness doesn't come bottled, but Stockholmers have a fascinating relationship with booze that you can investigate at the Spritmuseet (Museum of Spirits) on Djurgården Island. Stockholmers drink a lot of beer from the growing number of micro-breweries, but public drunkenness is isolated. Swedes rank about in the middle on the EU list of alcohol consumers.

The narrowest street in the city, Mårten Trotzigs Gränd, in Gamla Stan. Today, buskers make good use of the acoustics in parts of the Old Town. Music, pubs, restaurants and artisanal shops attract locals and tourists alike.

March is the climax of the skiing season, with the world's biggest cross-country ski race in Dalarna province. Not to be outdone, Stockholm puts on a race around Gamla Stan known as the Castle Sprint, trucking in snow if needed.

(Overleaf) You can't force swans to queue for lunch. The city provides daily lunches at Strömbro Bridge in winter when ice covers the birds' natural feeding sources.

APR

The cherry trees are blossoming in Kungsträdgården Park! The park's ice rink has been packed away and the strollers are back in force, cafés filled with Stockholmers grateful for spring sunshine.

APRIL IS A noisy mating month, at least for frogs.

By mid-April, the Japanese cherry trees (*Prunus accolade*) are blooming in Kungsträdgården Park. Cherry Blossom Day in late April is celebrated with the Japanese Association.

The park is hemmed by restaurants, cafés, a nightclub, galleries – and Sweden's richest banks, squatting in chunky granite. The copse of elm trees shading a café at the bridge end symbolises Stockholmers' environmental concern: in 1971, people rallied from all over the city to surround the trees and successfully save them from the axe. The underground Tunnelbana (Metro) station then being built to serve the park was given a different entrance.

The park's two main tree-lined parades are named for famous opera singers: soprano Birgit Nilsson and tenor Jussi Björling. They say Jussi used to duck around the corner to the Opera Bar for a drink in between acts at the Kungliga Operan (Royal Opera House). The trees sheltering the parades are small-leaved lindens (*Tilia cordata*), one of the few species that can take the rigours of Stockholm's climate and traffic.

Kids have been keeping an eye on the calendar. The Gröna Lund amusement park opens in late April. For parents and grandparents, a day at Gröna Lund will likely include the neighbouring ABBA Museum. Both are on Djurgården Island, accessible by bridge, bus and ferry. The island and a large adjoining sector make up the world's first urban national park.

Sometime in April: Kulturnatt – a night when zombies roam and music pulses in the clubs.

Full ferry services begin and about 40 vessels criss-cross the islands. A dozen old steamboats with engines serviced by volunteers manage the shorter routes. The steamship *Mariefred* and the steam sloop *Tärnan* are more than a century old. The steak and onions served on board used to be fried on the ship's boilers.

The ship, the *Vasa* – built for a warrior-king, never had a chance. Top-heavy with an extra, decorated deck, it tipped and sank before it could leave the port area. It was salvaged 333 years later, kept intact by mud and the low-saline Baltic Sea water, and is now housed in Sweden's most-visited museum. What was a fiasco has become a triumph.

(Overleaf)
Taking a break, leaning against the city architecture of three centuries, Stockholmers have forgotten winter. The island of Södermalm was once where smelly and noisy trades were exiled. The area is now the city's funkiest.

No 7.A-F

8
21

Almost through the gate! Locks and a weir between Lake Mälaren and the Baltic Sea straddle the Old Town island. The locks are for small boats, while the weir side can be used for white-water kayaking.

The Aula Medica auditorium at the Karolinska Institut. Almost half (40 per cent) of Sweden's medical research is through the Karolinska, which awards the Nobel Prize in Physiology or Medicine.

(Overleaf) The last day in April is Valborg Eve, celebrating a Christian saint but doing so with an ancient pagan ritual that welcomes fruitful May. Fires are scheduled at Skansen museum park overlooking the bay, and here, where the freshwater of Lake Mälaren laps the stones of the Old Town.

The longest ferry ride is from downtown to Arholma, the northernmost jetty in the archipelago. Students on holiday staff the cafeterias on board and provide nourishment for the four-hour trip. Bring a tent if you're travelling on a budget, otherwise check in at one of the two hostels. Arholma has a windmill, a secret fort, small farms – and a jetty dance floor in summer. The lovingly restored steam ferry, *Blidösund*, (p. 65) has music cruises all summer.

The archipelago has 24,000 islands. Sailors love it because the countless reefs are a challenge to navigate and the islands provide picturesque sheltered bays.

Sweden's national bird, the blackbird (*Turdus merula*), is back in town, sweet-talking the blackbird ladies from rooftops.

Socialdemokraterna

KUNGLIGA DRAMATISKA

Sun-hungry citizens take their seats at Dramaten (the Royal Dramatic Theatre), but to watch a different play. The building was financed almost entirely by the lottery and any surplus money was splurged on decoration and interior artwork. In the evenings, the gilt-bronze figures of Drama and Poetry usher in theatregoers.

(Overleaf) The swift waters between the lake and the sea are perfect for anglers. About 30 different kinds of fish thrive here, chiefly sea trout and salmon. Behind the bridge is the Ministry for Foreign Affairs.

FROM MAY UNTIL early October, sidewalk cafés spread out like forest grass, more per square metre than in Paris. And as cheap.

If it's a beaming May day, take a walk on Djurgården, the favourite city promenade. You're standing in the world's first urban national park, also known as Ekoparken. It used to be where royalty reposed in cane chairs to shoot at deer chased by beaters. Now, foxes, deer and hares scuttle around in the bushes while humans take the long path by the water.

Djurgården has everything: forest paths, canals and marinas, a sprawling outdoor folklore museum, cafés and restaurants, one of the world's great battleship museums, ferries and buses, and roller coaster rides.

The park's art museums are many: Thielska, the banker's legacy; Liljevalchs, with its spring salon; and the princely palace of Waldemarsudde, which was the home of a painter-prince, Eugen, who painted several of the City Hall's murals and mixed with artist friends and cronies.

Isbladskärret is a pond in the park that attracts birders from all over. Heron nest almost within earshot of traffic. Alongside the pond is a footpath called The Crown Princess Victoria and Prince Daniel Love Path, celebrating their popular public romance.

The blooming of the bird cherry (*Prunus padus*) and the lilac (*Syringa vulgaris*) is still news to be spread among friends. Now is the time to take a stroll around Skeppsholmen Island, or along the Monteliusvägen walkway on the South Island. Lilacs abound. And every park has at least one bird cherry.

In the summer season, Stockholmers seize the pavements for outdoor drinking and dining. The couple studying the menu are using two of the city's free bicycles to tour the Old Town.

(Overleaf) The Colonies. For a century, the city has provided plots of land for retirees and others. Once they were so far from the crowded streets that they were officially called *kolonilotter* ("colony allotments"). There are still around 45 sites, tucked away in corners of the city, with tens of thousands of individual plots.

Welcome
to Le Bar
please wait
to be seated
HILDA

ASKUNGEN
SJÖTAXI STOCKHOLM 08-661 88 80
www.josephine-charter.com
Josephine
GURLI
BILJETTER · TICKETS
NYBROCAFÉT

The ferries are longing for the islands but landlubbers have their own rewards in a sunshine stroll on upmarket Strandvägen.

(Overleaf) Take a walk on Monteliusvägen along the heights of Södermalm, the city's main island to the south. Bring some charcoal as well as something you can grill and there's a little park with a public barbecue. The walk is half a kilometre from end to end.

THE STOCKHOLM MARATHON is run in early June: two laps through streets and parks, egged on by onlookers and finishing in the old Olympic Stadium. In the 1912 Stockholm Olympics, a Japanese runner, Shizu Kanakuri, suffered mild heatstroke and stumbled into a private garden to rest. He could not finish the race. A hundred years later, his great-grandson ran a Kanakuri commemorative Stockholm marathon.

Midsummer Eve parties are welcoming occasions and you might find yourself with a bunch of Swedes, flowers in your hair. Myth says that seven wildflowers hidden under a pillow on Midsummer Eve will help you find a sweetheart. A basic Midsummer Eve meal should include: pickled herring, new potatoes, *snaps* (flavoured potato or grain spirit), hard bread and country cheese. Strawberries for dessert.

Trumpets and shouting in schoolyards, and truckloads of exhilarated school graduates cruising the streets.

School is out. The ritual involves unconventional transport from school to reception/party. The hired vehicles often have amplifiers blasting out music and carrying copious amounts of beer.

Before the truck ride, graduates wearing white peaked caps burst from the school's front door into a noisy crowd of proud friends and family.

Stockholm has about 90 high schools, including almost a dozen for special-needs pupils. Of graduates, over half go on to higher studies, more than in any of the other Nordic capitals.

Any time in June is high time to hop a ferry. You can go east to the archipelago or west up the flat valley, on the fresh water of Lake Mälaren. The major ferry quays are Blasieholmen, Nybrokajen and Stadshuset.

The quay at Stadshuset (City Hall), will lead you inland and in an hour you reach Drottningholm Palace. You don't have to be a royalist to appreciate the king and queen's cream 17th-century palace. It is the official residence of the Royal Family but you can tour parts of the building. The grounds include a baroque garden with gravel

National Day at the Skansen museum park isn't just for Stockholmers – people come from all over to celebrate the flag and the nation. The 6 June public holiday is a recent innovation and replaced one of Sweden's many Christian holidays.

(Overleaf) A bridge and a horde of marathoners dissect Långholmen Island, an idyll where prisoners were once held and executed. A ferry is puffing in from Drottningholm Island and its Royal Palace complex. The bridge, Västerbron, is one of 57 that link the city's islands.

KÄGELBANAN

Off in the distance, the Baltic Sea. Resisting the call of the archipelago, these revellers are delighting in a sprightly band, a cold beer and summer perfumes on the terrace at Mosebacke, the old theatre café. Close by are the restaurants, bars and basement comedy clubs of party-street Götgatan and hip "Sofo" (South of Folkungagatan).

The last of the coal-fired archipelago steamers is the *Blidösund*, offering live music all summer. In the background is the old Vaxholm fortress, only 60 minutes by ferry from downtown, first built in 1554 to repel enemies from the east. Rebuilt in the 19th century, it is now a museum and events centre.

paths, trimmed box (or boxwood) mini-hedges, fountains and a maze that no one could possibly get lost in. There is an open English park, much used by picnickers, and a Chinese Pavilion, originally built by a king for his queen in 1753. The Pavilion's fine rococo furnishings and chinoiseries indicate an historical link to Asia.

The court theatre is the oldest functioning baroque theatre in the world. Tickets for operas and music there are often snapped up a year in advance.

The entire Drottningholm complex – palace, theatre, pavilions and parks – is a UNESCO World Heritage Site. Sweden has 15 and three are in Stockholm; the two others are Skogskyrkogården (the Woodland Cemetery – see November) and Birka.

A two-hour ferry ride further up the lake archipelago will bring you to Birka, an ancient Viking trading centre. By the late 8th century,

traders from the Orient had found their way here, as well as Vikings from around Scandinavia.

The closest ferry trip in the other direction, seaward, is the 15-minute trip from Nybroplan, near Dramaten (the Royal Dramatic Theatre), to Fjärderholmerna, the "feather islets" named for the birds that flock there.

Take the ferry all the way north, the four-hour run to Arholma. Or all the way south to Landsort, or east to Sandhamn. Utö, Blidö, Muskö and Möja are all anxious for your appreciation. Venture out to the fisherfolk on Gillöga or Svenska Högarna, with their lighthouses and beacons. Everywhere, weather-smoothed rocks recall the latest Ice Age.

Theme ferries provide live music, all-you-can-eat shrimps, literary cruises. If you leave without having done a ferry trip, your Stockholm friends will be hurt. Some ferries are grand and fast, some are beloved museum pieces. All main routes are disabled-friendly, though not all the older boats.

Back in town, the farmers' markets open again on Saturdays, mid-mornings to mid-afternoons. No bargaining. The big market is on the south, Södermalm. The other caters to the posh clientele of Östermalm from a site in Tessin Park. Goat cheese, emu chorizos and root vegetables in various colours. At two end stations on the Metro line, Rinkeby and Skärholmen, what almost resemble Middle East markets are always busy.

Immigrants are gathered around the satellite communities along the Metro line. Back in town, each of Stockholm's four central residential sections has its own cachet. Few other cities define you, like Stockholm does, by your district: Södermalm is trendy. Östermalm is genteel. Vasastan is cosmopolitan chic. Kungsholmen is all of the above but less.

The Royal Palace of Drottningholm and the measured calm of its 17th-century French gardens, designed by court architects Nicodemus Tessin and his son (also Nicodemus), are recognised by UNESCO as a World Heritage Site. The palace has French, baroque and English gardens, a court theatre and a Chinese Pavilion, plus open meadows for picnic games.

(Overleaf)
The islands of Kastellholmen, with its turret, and Skeppsholmen. The larger island, once Sweden's naval headquarters, has several museums and an arts school. The smaller island is part of the Ekoparken, the world's first urban national park.

The Blue Gate (Blå Porten) is from the time when the island park of Djurgården was a fenced-in hunting preserve for royalty. Minutes from here are sedate restaurants, museums and the lush greenery of park meadows.

(Overleaf) The copper tents of Haga Park were built originally to house the royal guard, since King Gustaf III loved to spend time in the park, building a monument and politicking with friends and enemies. The lawn sweeps down behind here to the water of Brunnsviken Bay.

IF GOTHENBURG IS Sweden's industrial city, Uppsala an academic city, Malmö a city of immigrants, then Stockholm is a conglomeration of bureaucrats and creators, many in the IT sector. In July, even bureaucrats slow down and the parks are teeming.

Forty per cent of Stockholm's surface area is prioritised green and the municipality has seven nature reserves. A tree is planted for every new accommodation unit built.

Stroll through Haga Park for a waffle and whipped cream at the *koppartältet* (copper tent) built in 1790 for the mounted palace guard and a peek through the fence at Haga Palace.

Haga Park has romantic walks and grassy slopes and is Sweden's prime example of a classic, planned English park. A pavilion in the park was where a king, Gustaf III, rested before journeying into town to the Royal Opera, where he was assassinated. The murder, in 1792, was the basis for Verdi's opera, *Un ballo in maschera*.

Another spacious green area is Lill-Jansskogen, wild and hilly and great for joggers, many of them fit students from the Royal Institute of Technology. The campus stretches along one perimeter of the park. The closest Metro station is Tekniska Högskolan.

More than 60 outdoor barbecue spots with grills are waiting to be fired up. Most central is Ivar Lo's Park, on a height overlooking Riddarfjärden Bay and the city. And 15 minutes by public transport from Slussen's Tunnelbana (Metro) station is Hellasgården, Stockholm's largest activity centre, where they fire up the barbecue every Saturday and Sunday.

Or use your transport pass to go to Judarn Lake. Take the Metro green line to Åkeshov and step off the train into a nature reserve. The paths near the lake were formed 10,000 years ago by retreating Ice Age glaciers.

Picnickers fill the city parks on 26 July, Bellmansdagen, celebrating Carl Michael Bellman, an 18th-century minstrel-poet and a legendary carouser in Stockholm's parks and inns.

While Stockholmers are hitting the parks, tourists are filling the

The water at the top broadens into the Lake Mälaren delta, Sweden's ancient breadbasket. Klara Mälarstrand canal separates the island of Kungsholmen, to the left, from the posh Vasastan district.

(Overleaf)
The Wave Swinger at Gröna Lund amusement park.

Tens of thousands of small pleasure boats crowd the city's marinas and warm weather will see families going for a slow evening jaunt around the many islands.

(Overleaf) Clean water off Långholmen Island, within sight of the skyline. Industry re-zoning has brought back bottom-feeding fish and animals. Surface-swimming people too. The water we see is suitable for drinking.

(Pages 82–83) The island part of Djurgården, the urban national park. Forty per cent of Stockholm is a green zone. The Nordic Museum (Nordiska Museet) dominates the foreground with the Vasa Museum under rigging to the right. The low, flat building in the right foreground is the Junibacken museum centre for children.

museums. From the wall-to-ceiling windows at the Moderna Museet café/restaurant, you will be looking straight at the Vasamuseet (Vasa Museum) with its giant mast. Behind it, the Nordiska Museet (Nordic Museum), built in northern European Renaissance style. The main hall is the largest covered space in Sweden except for cathedrals and Stockholm's Globe Arena. Not far from the Nordiska Museet is the ABBA Museum. One of its features is a telephone line linked to Benny's studio.

Just across the harbour, but out of sight, is Fotografiska, the photographic museum in an old harbour building. Close to it is the Stadsmuseet (City Museum), and you will have already passed Nationalmuseum (Museum of Fine Arts) on the other side of the bridge.

At the end of the month or the beginning of August, the annual Pride Festival week floods the streets with LGBT supporters from near and far.

A UN Habitat Report praises Stockholm for "fostering social inclusion", noting the "removal of barriers to full engagement of women, youths, the homeless, the elderly and the disabled." Stockholm scores high in democracy as well – voter turnout for regional and local elections can top 80 per cent, while the other Nordic capitals struggle to reach 60 per cent.

July is the slowest month for the property market, which crescendos in late spring and mid-autumn. The city is composed of a metro area and a city district, with an exclusive inner city defined by a ring of five old customs posts. Inside the ring, rents and property prices rise.

The city's economy weathered the storms of the 2008/9 global financial crisis but property is expensive relative to the cost of living. The city ranks in the top 50 of the world's 780 costliest places to live. And in the top third for expensive alcohol and tobacco.

epic

VIKING LINE

AUG

The Pride Parade
in August is
Scandinavia's
largest gay festival.
Stockholm has no
identifiable gay
district, largely
because the LGBT
community is
relatively integrated.

THE EXPLOSION IN bicycle traffic has led to multiple bike lanes and free-share bikes at almost 100 stations. Use your charge card or the Access public transport card. Stockholm also has 100 Tunnelbana (Metro) stations, most with artistic decoration. The T-bana is proudly called "the world's longest art gallery".

A congestion tax levied in the downtown zone has kept the city free of traffic jams, except at the entrance points. Of the Nordic capitals, only Copenhagen has fewer cars in relation to population. Traffic in Stockholm is sedate, and safer than in the other major Swedish cities. (Sweden has the lowest incidence of traffic deaths in relation to population in all of Europe.)

Sveavägen is Stockholm's major boulevard, originally intended to be as long and as wide as the Champs-Élysées in Paris. On a warm Friday or Saturday evening, you might hear the purring parade of restored American cars from the last century. At the wheels are the curators of the golden age of American automobile. Grizzled gentlemen, once rebels and now comfortable suburbanites, flaunt their affection for a lost age.

In mid-month, the Royal Philharmonic Orchestra performs outdoors at the Sjöhistoriska Museet (Maritime Museum), with a backdrop of forest and water.

School is back. Pupils returning cheerful or reluctant. The city is slowly getting back to its pre-summer mode. This might be an opportune time to take a longer ferry trip. The massive Baltic ferries are synonymous with partying.

The Finland ferry leaves early evening, arriving in Helsinki in the morning. You get the day ashore then depart for Stockholm in the evening. Since Sweden and Finland are both in the EU, international waters' duty-free rules no longer apply, but many ferries solve the problem by docking briefly in Åland, a group of islands with semi-autonomous status and thus other rules. Other popular routes are to Estonia, Latvia, Poland, Denmark and Germany.

Skies darken but the multi-purpose Ericsson Globe arena is a beacon to the southeast. The central business district is in the foreground; the Royal Palace and the Old Town are mid-picture; and to the lower right are the five white-sided office buildings called *trumpetstötar* ("trumpet blasts") by a proud city council in the optimistic 1950s.

(Overleaf) Hot air and advertising. The balloons are back, giving wedding groups and sightseers a loving look at the city. Stockholm is one of few major European cities that permit intrusive balloons over a built-up area.

VINGA

Codan

This is the last month of high jinks in the city. The vintage flea market on Hornstull Strand is crowded, food trucks zip around, barely stopping to serve diners. There are fancy dress fun runs on Söder Island, some in the middle of the night, and Kungsträdgården Park invariably has a market or a show.

Life is good if you're a moose in a living museum. The Skansen museum park has examples of all of Sweden's animals and trees from throughout the Nordic region. The King of the Forest expects the best.

Protecting the southside Södermalm district like a fortress, the residential buildings of Mariaberget catch the morning sun.

(Overleaf)
The Royal Philharmonic Orchestra performs in Djurgården Park every August.

(Pages 94–95)
Ferry traffic is intense. Goods and people criss-cross the Baltic to Finland, Estonia, Latvia, Lithuania, Germany and Denmark. Restrictions on fuel type and speed help protect the inlet waters and bird life.

Filharmonikerna i det gröna
DN
DN DN
Parkteater
SF SKANDINAVISK FRAKT AB
ICM
DN
Parkteater

BIRKA PARADISE
MARIEHAMN
GABRIELLA

VIKING LINE
BEWE
1
2
3

The Konserthuset (Concert Hall), with its sturdy Corinthian columns, hosts classical and jazz concerts as well as the Nobel Prize ceremony. The fruit and vegetable stalls, where bargaining is standard, are in the old haymarket.

(Overleaf) Kayaking in Brunnsviken Bay. To the left, the Great Lawn sweeps uphill to the pavilions and palaces of Haga Park, part of the national urban park. Hidden in the dense trees to the right is the Gustaf III Pavilion where a murdered king spent his happiest days.

SEP

Autumn mist turns Gröna Lund amusement park into a fantasy city. To the extreme right, on Kastellholmen Islet, is a fortress turret.

THIS IS THE perfect month to follow the gravel path around Brunnsviken Bay in Haga Park. From the city's northern customs gate at the tip of Stockholm's busiest boulevard, you enter an English park and follow the northern waterline to the Bergianska Conservatory. On your way you pass century-old wooden villas formerly owned by artists. Rent a canoe and paddle around the bay.

This is peak season for wild mushrooms, or fungi as they're more correctly but less appetisingly called. The influx of foreign nationals, remembering *almost* identical fungi from their countries of origin, has brought an increase in poisoning accidents. But mushroom-pickers are generally an astute lot and countless illustrated books have been produced on the dos and don'ts of gathering and eating. Mushroom picking is a golden opportunity for Stockholmers to regenerate, or invent, countryside memories, and by mid-October, the woods at the outskirts of the city will have been picked clean of chanterelles (*Cantharellus ciabrius*).

At the vegetable market at Hötorget, the stalls are packed high with chanterelles imported from across the Baltic, although the mainly Middle Eastern stall-owners will swear they are from pristine Swedish locations. Hötorget is the single place in town where you can bargain with traders, but be aware you're tangling with masters of the art.

September is a sports month, with track and field culminating and football competitions heating up as the temperature drops. Ice hockey begins in September and lasts until May, with a break at Christmas.

The Stockholm Whitewater Games is a kayak slalom and kayak cross competition under the Royal Palace windows. In kayak cross, contestants are allowed to shove and tackle each other. If that's not rough enough, the organisers also arrange canoe polo – another way to get wet in a kayak.

Ice hockey season begins. See a game at the Ericsson Globe. You can barely see the puck from the seats in the third tier but the acoustics are perfect. The roaring hometown fans take full advantage. If you're not a hockey fan, you can see the Globe from the outside and from the top with the Skyview ride.

The Globe is used to represent the sun in the world's largest permanent scale model of the solar system. Other buildings, representing the planets on the same scale, are scattered across Stockholm and the rest of Sweden. "Saturn" is in Uppsala, "Neptune" is Söderhamn on the north-central coast and "Terminal Shock", the edge of the heliosphere and the furthest place solar winds reach, is Kiruna in the far north.

The last day of the month is Hamnens Dag (Harbour Day). Dozens of vessels tie up at the two major downtown quays at Stadsgården and Skeppsbron. Open for inspection are coal-fired steamer

Changing guard at the Royal Palace in 17th-century uniform. The original guards, founded in 1523, doubled as firemen.

A walk around the periphery of Djurgården Park, remembering summer in the archipelago. The ferry is headed for the pier. The building with a tower is an old people's home. It is claimed that a visiting naval ship mistook it for the Royal Palace and fired a salute.

STOCKHOLM

Fotografiska
BUSS

The old customs house now hosts Fotografiska, a successful private photographic museum in Stockholm's dock precinct.

Stockholmers have always loved the *konditori* or cake shop. Here, Oscar's Café on Narvavägen.

(Overleaf) Four young people, dressed to amaze, put stars in the eyes of millions around the world. The ABBA Museum is a celebration of one of the music world's greatest success stories.

ferries, old tramp freighters, police boats, coastguard surveillance vessels, training ships and barges.

For private boat owners are about to haul their boats out of the water. In multiple marinas around town, boats are placed on stands or taken to storage. The autumn marks the end of the night-watch period, when members commit to at least one all-night shift.

The end of September also ends any real hope of hot sunshine. But it's an exciting time for a Stockholmer's brain – and dancing feet. There are lectures at the handsome Stadsbiblioteket (City Library), and rock music is going back underground to the clubs and cellars on Södermalm island to ferment happily. The big acts have filled the stadiums in the summer, now lesser-known new musicians and shows get their chance to woo Stockholmers. The Cirkus arena on Djurgården Island and Konserthuset (Concert Hall) in town are hot venues.

greatest hits
ABBA
WATERLOO
ABBA

A renaissance in cycling has changed the character of Stockholm's work commute. The city's Accessibility Strategy is based on more buses and bikes, to deal with the city's swift growth. Come winter, many cyclists simply change to studded tyres.

4 OCTOBER IS CINNAMON Bun Day, which is good news for the city's 500-plus cafés and cake shops, although the bun isn't specifically a Stockholm thing. But the city's cafés are vibrant and the baristas skilful. Coffee has always been served strong in this part of the world, and the world's top six coffee-drinking countries, in terms of volume, including all the Nordic nationalities plus the Dutch.

Stockholm is on the world coffee appreciation map with more baristas and confectioners recognised internationally. Try a *bulle* (bun) with your coffee. There are several kinds, but don't leave town without having had a cinnamon *kanelbulle* or a cardamom bun. If your sweet tooth is even sweeter, go for a *dammsugare* ("vacuum cleaner") at Vetekatten, one of the oldest traditional cafés.

Stockholm's hippest restaurants and ethic-cuisine mix is on Södermalm. The island is studded with New York and Continental influences, and eateries that don't cost too much. Food trucks – mobile gourmet fast food – prowl in daytime.

The city used to have a hot dog culture, but the sector lost customers first to McDonald's and then to the Thai cooks who have now taken over the fast food kiosks. If you do find a *varmkorv* (hot dog) kiosk, order the *trålare* (trawler). For a reasonable price you will be served a hot dog stuck in chopped shrimp and mayonnaise and wrapped in thin, supple Norrland bread. The most indigenous outdoor fast food is the fried Baltic herring on hard bread, but it can be hard to find. Order *strömming* in a restaurant instead, fried or pickled.

From 1 October to 15 April, motorists are permitted to change to studded tyres. From 1 December to 31 March every car must have *some* type of winter/snow tyre. Studs can be lifesavers on black ice but are banned on some densely populated streets because of particle dissemination. Overall, the city's air meets EU directives for the protection of human health, and air quality has been improving for decades.

One of the city's shortest ferry lines shuttles between the city and Djurgården Park. To the right, the Nationalmuseum (Museum of Fine Arts), one of Europe's oldest art museums, founded in 1792, by King Gustaf III, once resident in the Royal Palace right across the water.

STOCKHOLMS STRÖM 2

Stockholm has 29 parishes and dozens of old churches. Maria Magdalena kyrka, autumn leaves now swept up and tidied away, sits peacefully amid the hum of Hornsgatan. Across from here is an artistic district among old-fashioned cobbled streets.

(Overleaf)
The food hall at Östermalmstorg. Several inner city squares have food halls, with the best-hung meat and game, cheeses with difficult names and gourmet lunch counters.

(Pages 116–117)
Skeppsbron is where the merchants built their palatial offices for import and export. The buildings to the left were best positioned to spot the incoming sailing ships laden with goods. The massive Royal Palace dominates the north side of Gamla Stan, the Old Town.

The Stockholm Jazz Festival is often held in October, and at several venues across the city.

Evening education courses kick in. Stockholm's working class has become a service-executive class but the old social democratic tradition of evening education survives and thrives. What's on offer? Learn how to felt wool (soap, wool and hot water are involved), sew lampshades, be a better movie extra; or do magic tricks (bring your own coins, cups, spoons, balls and wear a long-sleeved sweater or jacket with pockets).

Evening lectures are also starting. Some of the best are in the huge hall of the City Library. The library was designed by Gunnar Asplund, an early 20th-century architectural trendsetter. It was the country's first self-service library.

The inner city buildings are coloured various shades of yellow, ochre, beige, amber, apricot, lion, maize, cream, ecru, jonquil, khaki and buff – in short: brown, but good-looking brown. The lighter shades help reflect light in the murk of winter.

By now, joggers are out in force. October is also when the hardcore runners take off into the forests in Sweden's very own invented sport of orienteering. The four major orienteering events are racing, ski-orienteering, mountain-bike orienteering and precision. It's the perfect way to get lost in the countryside then find your way home. To make it more difficult, the Enskede club in south Stockholm has its night-time Natti-Natti orienteering event this month.

The floorball season begins. This is a kind of floor hockey played with hockey sticks on a plastic-mesh hitting surface and using a dimpled, hollow plastic ball with holes in it. Another game invented in Sweden, in the 1970s.

And look, there's a string of kids herded to a park. Their high-visibility vests have the daycare's name and telephone number. Stockholm kids are kept outdoors as much as possible.

In late October, many archipelago islands celebrate *eld över hav och land* ("fire on sea and land"). Island inns fire up their barbecue grills and ferries shuttle back and forth.

68
HUSMANS DELI
67
HUSMANS DELI
HUSMANS DELI

LISA
LISA ELMQVIST
SMØRREBRØD
NYBROE

HOTELL REISEN

A daycare group wearing hi-vis vests prepares to move off. This is Humlegården Park, named for an optimistic hop farm in the 16th century. Obscured here by vegetation is the statue of Carl Linnaeus, the Stockholmer who invented the system by which the world names and classifies its flora and fauna. The backdrop is the Royal Library, a stately building on top of multiple underground levels of carefully preserved books and papers.

IT'S GETTING DARKER. Stockholmers argue whether November or February is the most crushing month. Light is longed for. City planners years ago abandoned a standard lighting fixture and local lighting styles are part of local character. Check out the different street lights at these bridges: Djurgårdsbron, Vasabron, Norrbro – and at Norrmalmstorg and the nearby Dramaten (the Royal Dramatic Theatre).

All Saints/All Souls Day at the beginning of the month is marked at Skogskyrkogården (the Woodland Cemetery) with tiny flares scattered through the pines to remember ancestors and loved ones. The cemetery is a UNESCO World Heritage Site.

By mid-month, all dead leaves have been vacuumed from the inner-city parks. The first snow has fallen. Mid-November is when Christmas trees pop up on city squares.

The approaching winter demands fitness of the citizens. A classic power-walk is along the waterside path at Norr Mälarstrand, a favourite address for movie stars in the 1940s. The path leads you gently from City Hall along Norr Mälarstrand to a park and Västerbro Bridge. Across the water is Långholmen Island with the old prison, which is now a hostel for those on good behaviour. You can continue around Kungsholmen Island but that's 9 kilometres. Walk around Långholmen instead: 70 minutes at normal pace.

Getting through the long winter calls for health and fitness, but since more time indoors becomes inevitable as the temperature drops, Stockholmers' thoughts turn to home improvement. The Stockholm International Trade Fair complex at Älvsjö, 10 minutes by train from Central, is pulling in crowds to its autumn fairs themed to kitchens, food and health.

The Stockholm Film Festival chooses November because of that pull to be indoors from cruel, gusty rain. Stockholmers love to come in from the cold and bury themselves in popcorn in a cinema seat. Winter is also when the Stockholm brain enters intellectual mode. Art films get their biggest audiences, either to the multiplexes in

City Hall. A happy marriage of Nordic and Venetian architecture. Eight million bricks were used. The columned arcade gives onto a cobbled courtyard. The best artists and craftsmen of the early 1900s beautified the interiors.

Heron City, Kista, Råsunda, Vällingby, Sickla and the old studio complex, Gamla Filmstaden.

Back outdoors, thriller writer Stieg Larsson's *Millennium* trilogy is celebrated in a *Millennium* Walk, taking readers to sites that recall the dark deeds of Lisbeth and Mikael. November is perfect for this.

30 November is Cabbage Dolmades Day, celebrating Sweden's version of Middle Eastern stuffed vine leaves. The dish came to Sweden after the defeat of the warrior-king, Charles XII, in what is now Moldova.

November is the month when fewest children are born in Sweden, which probably says more about February than November.

Muffled pedestrians and slower traffic change the dynamics of the city when the snow comes. Stockholm can handle snow and no one gripes.

All Souls Day. Stockholmers may be secular but they save great respect for their cemeteries, especially on this day. Not even one in five Swedes believes "there is a God". In Stockholm, the influx of non-Christian religions may confuse the overall picture.

(Overleaf) The city's prime real estate street, Strandvägen, was built by adding reclaimed land to what was once a semi-rural slum area. Upper right is Djurgården Park. The Kaknäs communications tower is busy receiving and transmitting signals.

STOCKHOLM

DJURGÅRDEN 7

avenova
LEIMDÖRFER
Telia
savills
Europcar
HYR
HOS OSS

DEC

The Obelisk and its fountain form the city hub of Sergelstorg. Underneath the square – which is in fact a superellipse – is a beehive of shops and tunnels to more shops, buzzing with Christmas excitement.

(Overleaf)
The fanciest feast of the year. The Nobel banquet at City Hall under the exposed brick walls of architect Ragnar Östberg. The menu and manners are impeccable. The Prize's founder, Alfred Nobel (1833–96), was a man of fine tastes and idealistic dreams.

THE FIRST BLIZZARD. Public transport delays. Didn't the transport business see winter coming? Rail commuters are hard hit but the city's basic transport infrastructure is second to few in the world.

It's a month of excitements: the holidays are closing in. And early in the month, the city enjoys its favourite formal moment – Nobel Day on 10 December. (Alfred Nobel, the gentle inventor of dynamite, grew up in St. Petersburg and died in San Remo but was born in Stockholm and as a small boy sold matches at Norrmalmstorg Square to boost family income.)

For days now, the Grand Hôtel and all the other leading hotels have been checking in the world's most eminent scientists, economists and one or two writers. The awards ceremony is formality at its old-fashioned best, involving a monarch and starched shirts, white tie and tails, and gowns.

The six Nobel Prizes include one that wasn't on Alfred's original list, and one that isn't awarded in Stockholm. The Peace Prize is presented in Oslo by the Norwegian parliament because Sweden and Norway were united when Alfred died in 1896. The Memorial Prize in Economic Sciences was not on Alfred Nobel's list but the wealthy Nobel Foundation and the equally wealthy Riksbank (Bank of Sweden), accepting that economics has become one of our most vital sciences, added the award. The prizes bring Stockholm incalculable prestige.

That evening, there's a dignified banquet at City Hall. It's the élite of knowledge dining with the worthiest local citizens, and everyone is looking their best, honouring the inventor who left his fortune to hope and knowledge.

The City Hall's tapering tower is often used as the model for gingerbread houses baked in family kitchens in the run-up to Christmas. The Architecture and Design Centre has a yearly gingerbread house contest.

Stockholm is strongly secular. Its churches are lovingly preserved but largely empty. However, the four Sundays of Advent

The story of St. Lucia is a long-forgotten tragedy – for Stockholmers, only joy remains. The Luciafest on 13 December is marked by much respect for tradition, as well as some pure partying. The day begins with candles at dawn and *pepparkakor* biscuits and coffee.

(Overleaf) Fill up on cheese, sweet *glögg* wine and hot dogs while shopping for Christmas. The four Advent Sundays draw present-hunters to Gamla Stan, the Old Town, where country handicrafts delight the city folk.

Keramik

have a compassionate effect on hustling, bustling Stockholmers. The lights and candles in windows illuminate our immediate surroundings.

This is the time to think of presents. Christmas markets are set up in Kungsträdgården Park and Stortorget Square in Gamla Stan. Skansen, the museum park, has an elaborate one among the reconstructed historical buildings. At winter solstice (21 December) Skansen also has an ice-and-fire show.

It's cold. The city's pigeons huddle for warmth. The swans and archipelago birds at Strömbro Bridge are getting cranky, waiting to be fed. Stockholmers, those who are fed and those who are huddling, are looking forward to January.

A halo display, sometimes called a "sun dog". This atmospheric optical phenomenon was recorded in the oldest colour painting of Stockholm, from 1535.

The store window to see at Christmas is at the NK department store on Hamngatan, across from the Kungsträdgården Park. The display might have bearded trolls, reindeer and moose, interacting cleverly with merchandise.

(Overleaf) The ship is the full-rigger *af Chapman*, once a cargo ship, now a hostel. The square lines of the Royal Palace dominate the centre skyline. It is a city dominated by light (in winter by decorative lighting), which makes the sudden clear skies of a winter day a happy bonus.

Facts about Stockholm

Stockholm is the capital of Sweden and the country's largest city. It is the country's cultural, media, economic and political centre. The municipality has 900,000 inhabitants while over two million people inhabit metro Stockholm, making the city the biggest urban conglomeration in the Nordic area. The number of passengers using the Metro each day is 680,000.

Almost a quarter of all Sweden's business companies are in the Stockholm area. Stockholm is also the Swedish city visited by most international guests, with over one million foreign visitors a year.

The city was founded in the early 13th century. Its location, on islands where the Lake Mälaren delta spills into the Baltic Sea, on the border between the provinces of Uppland and Södermanland, was strategic for both trade and defence, strongly contributing to Stockholm's development as a mainstay of Sweden's defence and the country's largest and most important trading centre.

The oldest buildings in the inner city are the cathedral of Storkyrkan in Gamla Stan and the church on Riddarholmen, which date from the late 13th century.

Stockholm was first mentioned in written records in 1252 and was accorded city privileges by royal decree on 1 May 1436, now accepted as the date when Stockholm became Sweden's capital. Its role as administrative centre and official capital was confirmed in the constitution of 1634 when the monarchy, parliament and central administration were established in the city.

In east Stockholm, the widespread archipelago begins, its 30,000 islands forming one of the world's most beautiful island accumulations.

One third of the city area is park or greenbelt.

In 1998 Stockholm Municipality was named European Capital of Culture and in 2010 it was European Green Capital.

Stockholm in figures

Area, division and land

Land area, km^2	188
thereof parks and green zones, %	33
Population density per km^2	4,687
Numbers of municipalities	1
Numbers of parishes	27

Population 31 Dec each year

1980	647,214
1990	674,452
2000	750,348
2012	881,235
2022 (predicted)	1,010,493
2030 (predicted)	1,094,510
Share of the country (2012), %	9.2

Age distribution (2012), %

<19 years	21.3
20–44 years	40.8
45–64 years	23.6
65< years	14.2

Mean age of mothers for first child	30.5 years
Mean age of fathers for first child	32.9 years
Average number of children per family	1.8

Average age	38.3 years
Average longevity for men	81.1 years
Average longevity for women	84.6 years

Foreign-born persons

		%
Finland	17,579	8.7
Iraq	16,374	8.1
Iran	11,429	5.7
Poland	10,612	5.3
Turkey	7,429	3.7
Somalia	7,364	3.6
Chile	5,440	2.7
Germany	4,791	2.4
Ex-Yugoslavia	4,785	2.4
China	4,746	2.4
Other	111,278	55
Total	201,827	100

Percentage of population with college education (25–64 years)	56 %

Mean income (2011), in kronor

Men	372,500
20–64 years	382,400
65< years	321,800
Women	277,900
20–64 years	296,100
65< years	211,700

Taxes

Total per 100 SEK	29.65
Municipal taxes per 100 SEK	17.48
County council taxes per 100 SEK	12.10
Funeral fee per 100 SEK	0.07

Culture 2012

Theatres (2008)	53
Music- and dance venues (2008)	41
Museums	70
Art galleries (2011)	7
Public Libraries (2013)	51
Cinemas	77

The highest buildings

Kista Science Tower	158 m (518 feet)
Kaknäs Tower	155 m (508 feet)
Victoria Tower	120 m (394 feet)
Klara Church	116 m (381 feet)
City Hall	106 m (348 feet)

Weather 2012

Mean temperature (°C)	7.2
Max temperature (°C)	26.9
Min temperature (°C)	-16.5
Precipitation (mm)	779
Precipitation days	199
Hours of sunshine	1776

Sources
City of Stockholm, SCB, SMHI and Sweco

Picture Sources

Jacket, front and back Jeppe Wikström

Erik Abel/TT 84

Sören Andersson/mediabank.visitstockholm.com 12

Anders Blomqvist/Getty Images 90

Rolf Christensen/TT 96

Malcolm Hanes 24–25

Leif R Jansson/TT 108

Peter Lydén/TT 123

Henrik Montgomery/TT 128–129

Tobbe Nilsson 76–77

Max Plunger 104

Fredrik Schlyter/Naturbild 33

Maja Suslin/TT 62–63

Erik G Svensson 13, 17, 18–19, 26, 38, 45, 46–47, 48, 64, 70, 72–73, 80–81, 94–95, 112, 114–115, 121, 122, 126, 132–133, 134, 136–137

Henrik Trygg/mediabank.visitstockholm.com 28–29, 35, 131

Henrik Trygg/Naturbild 135

Roger Turesson 34

Jeppe Wikström 1, 4, 7, 8, 10–11, 14–15, 20, 23, 30, 36–37, 41, 42–43, 44, 50–51, 53, 54–55, 56, 58–59, 61, 65, 67, 68–69, 75, 78, 82–83, 87, 88–89, 91, 92–93, 98–99, 100, 102, 103, 105, 106–107, 111, 116–117, 118, 124–125